Powell Perry

Manual of practical photography

A guide to the reproductive processes

Powell Perry

Manual of practical photography
A guide to the reproductive processes

ISBN/EAN: 9783743378278

Manufactured in Europe, USA, Canada, Australia, Japa

Cover: Foto ©berggeist007 / pixelio.de

Manufactured and distributed by brebook publishing software
(www.brebook.com)

Powell Perry

Manual of practical photography

Uncle Alberts

MANUAL of PRACTICAL

PHOTOGRAPHY

and

Guide to the Reproductive Processes

Posthumously Published by his Dutiful Nephew

Uncle Albert

Printed and Published by Perry Colourprint Ltd., London, S.W. 15.

Foreword by the Author's Nephew

Obviously one's fellow men are not to be trusted ! The torrent of text books on THE ART OF PHOTOGRAPHY that has appeared during the past fifty years or so bears eloquent, if dumb, witness to this unpleasant truism : for how, otherwise, could so many have based, like crawling parasites, their spurious writings on the lovingly garnered information and painstakingly original research of my Uncle Albert ? It is indeed ironic that the one work of his that was *never* published should have been so brazenly pilfered—in embryo, as it were—whilst, to the best of my knowledge, *nobody* has ever dared to quote as much as a single phrase from any of his forty-seven volumes of published treatises on subjects ranging from " The possibility of a study of Amoebae as an introduction " to simple division in junior schools " to " Stamp collecting in North Borneo."

Uncle Albert collecting masses of data by comparing his own density with that of a block of granite. The thoroughness with which he entered into discouraging experimental work of this kind, even at an advanced age, is truly indicative of that rugged persistence which is the earmark of the sincere seeker after knowledge.

Perhaps it is inevitable that one so sweepingly versatile as my Uncle Albert should have been a little garrulous. Perhaps, too, the medicinal spirits that he had recourse to as a stimulant after long hours in dark rooms served to loosen his tongue as well as to " fix his collar down " . . . (pun, *collodion*—ED.) . . . as he was wont, jocosely, to remark.

However, it is not with the *causes* but with the *effects* of such indiscriminate confidences that I am here concerned—effects, the very existence of which serve to indict far more effectively than any reproofs of mine *the vicious practice of literary and scientific plagiarism.*

I ask you, dear reader, to examine any six text books on photography, chosen at random from the " P " section of your local Public Library : What do you find ? The most casual examination will suffice to prove that every writer says precisely the same thing in precisely the same way. If you persevere and read three or four pages of each book thoroughly you will find the same chemicals mentioned, the same methods of handling detailed and the same results arrived at. Six times you will read that a good developer can be made from :

Saturated solution of ferrous sulphate ..	2 ozs.
Glacial acetic acid ..	$\frac{1}{4}$ oz.
Alcohol	1 oz.
Water	16 ozs.

and six times you will be told that :

$$3 \, AgNO_3 \quad 3 \, Fe \, So_4 = 3 \, Ag - Fe_2 \, (So_4)_3 - Fe_2 \, (No_3)_3$$

Millions of words and acres of paper wasted on unimaginative repetition—what better proof than this could there be of the utter sterility of scientific cribbing?

No ! Uncle Albert's dependents may have been robbed of some of the posthumous fruits of his scientific and artistic labours, but, in presenting his " Manual of Practical Photography, etc.," to the public, I am happily conscious that not only is a belated recognition being accorded to original research of a high order, but a blow has been struck that will help, in some measure, to cleanse the Aegean stables of photographic upstartism.

———

In conclusion I would like to stress that this work must, by its very nature, be more of a spontaneous personal record than an exhaustive and ordered treatise. As one contemporaty critic happily puts it . . . "One of the many things that " ' Uncle Albert's Manual, etc.,' has in common with the ' Notebooks ' of " Leonardo de Vinci is an eclectic discursiveness that takes merely technical " difficulties in its stride."

Indoor Photography.

It is gratifying to remember that the modern (sic. 1890—Ed.) Art of Indoor Photography has its feet very firmly planted on the pictorial achievement of the past. The very phrase "Necks please!"—beloved of the busy commercial photographer—is a quaint survival that can be traced right back to where the Pre-Raphaelites started from. An appreciation of this historic fact has prompted the keen photographic Artist to affect the velvety looseness of dress and abundance of hair that is to-day recognised as the distinctive uniform of pictorial genius. However, the hair should not be worn so long that it hangs over the camera lens as well as the collar : it has been found that only the very best photographers have that innate flair for composition which enables them to work strands of hair into the subjects in a natural sort of way

Apropos of INDOOR PHOTOGRAPHY, and particularly *Portraiture*, I cannot entirely agree with W. J. Loftie who, in his book "A Plea for Art in the House," says : . . . "Photography is of little use for portraiture. I mean that large pictures "of landscapes in photography are much more common and more pleasing than "large likenesses. The vulgar staring portraits produced by many photographers "do not bear enlargement. . . ." To this type of irresponsible criticism one can only respond that the *vulgar staring* is an attribute of the subject and not of the photographer, who usually prefers peep discreetly from underneath a black cloth. Photographers, like their brothers in Art, cannot always be choosers, and one might with equal justification indict Rembrandt for picturing the *vulgar staring* Syndics of the Cloth Guild. Anyway the point can be avoided by concentrating on profiles and using a nice soft focus lens. Indeed, in this way it is possible to satisfy both of Mr. Loftie's objections as I have myself by these methods produced portraits which, from a short distance, are quite indistinguishable from landscapes of the popular "Mist in the Highlands" genre. DOUBLE-SUBJECT photography of this type—combining, as it does, the universal charm of landscape with the strong personal appeal of portraiture—is worthy of the attention of all progressive photographers.

AN OUTSTANDING EXAMPLE OF INDOOR EXPOSURE.

—In this informal get-together a new student is shown toasting senior members of my " Anti-Under-Development Class." The appearance of spontaneous gaiety is entirely illusory, since this particular exposure went on for about four hours (not including two ten-minute intervals) and both toaster and toastees had to ease their elastic boots several times.

NECESSARY EQUIPMENT.

Nowadays the virtues of both the HAND-CAMERA and the STAND-CAMERA have been combined in a popular *all-purpose* or " HAND-STAND " Model. Lest this description should tend to mislead the unwary amateur I hasten to assure him (or her—? ED.) that the phrase " Hand-Stand " is usually regarded as applying to the *Camera* and *not* to the attitude to be adopted by the photographer, or, except under special circumstances, of the subject. It has been found, however, that if the subject flatly refuses to respond to the usual invitation to " Watch " for the Dickey," (*see chapter on " Exposure* ") the photographer can, as a last resort, usually command some degree of attention if he (or she—? ED.), suddenly stands on the hands (presumably on his, or her, own hands—ED.), or, better still, hangs upside down from a trapeze, gasolier, or other convenient swinging fixture. I recently came across some interesting if rather involved statistics on the subject, which, boiled down, prove that, all things being equal, the degree of attention commanded by the photographer reversing his usual position varies in inverse ratio with the age, sex, blood pressure, underwear and general susceptibility of both the photographer and his (or her—ED.) subject. In passing it is interesting to remember that the standard CAMERA TRIPOD was designed by the ingenious inventor of the common or garden (or park or beach—ED.) Folding deck chair, and suffers from many of the whimsical aberrations of its prototype. Once the contraption is opened up, and the wounds on the fingers have healed, it is better to forget that it *is* possible—at risk of limb, temper and time—to make it fold up again into a neat (*see maker's catalogue*) bundle only about four-fifths the size of the fully extended tripod.

If you *must* use a tripod the model shown here, although practically useless, has the advantage of folding up into a convenient bundle about the size of an umbrella ; *I* prefer to carry an umbrella.

Apropos of Double-subject photography... what do you think of this photopicture entitled "Washer girls! *How are you?" In a sense this is a multiple subject, for, in addition to the figurative interest of the foreground, the rather uninspiring mass of foliage in the background is cleverly broken up with gay little patches of undies. The fact that somebody's mother isn't using a well-known patent washing powder is also rather obvious.*

However, it is only fair to say that the best indoor work is done on the stand, and not with the hand camera. Supplementary equipment should include a large aspidestra or two, a few plain and fluted columns with removable bases and capitals (it is the practice, presumably to avoid the risk of contravening the Law of Copyright, always to combine the Four Orders of Architecture in single composite columns for photographic purposes—the calibre of an Indoor Photographer can to a large extent be measured by the ingenuity with which he does this). A Jacobean Jardiniere, several hundred yards of hard-wearing drapes, a painted backcloth showing the interior of the Main Banqueting Hall at the Palace of Varieties (Versailles—? Ed.) an assortment of false moustaches and toupees, and a glass of water, complete the standard studio equipment.

It is not considered correct to solicit customers by standing invitingly outside the door of the studio and whistling or beckoning with one or more fingers. A discreet notice with perhaps a few chastely framed examples of photographic portraiture and landscape is deemed to be dignified and sufficient. The swinging sign shown on page 12 struck just the right note.

Once inside the subjects must be made to feel at ease and encouraged not to stare at the camera as if expecting it to leap at them and grasp them by the throat; some photographers playfully pat the camera, just to show that it won't bite. Don't overdo this—a tap is sufficient and should not be followed by a hearty kick or resounding thump. conduct of this kind only serves to alarm the client and doesn't do the camera any good.

If two people enter the studio at the same time it is wrong automatically to assume that they wish to be taken together with one seated and one standing with the right-hand on the other's shoulder, against a background of ruined pergolas . . . they probably do, but it is only courtesy to enquire. Composite photographs of total strangers do not, as a rule, sell well (for exceptions see chapter on TRICK PHOTOGRAPHY.)

We quote from a contemporary suggesting that " . . . . When persons are about " to have their portrait taken, they should, if they wish to secure the most perfect " resemblance of themselves as they generally appear, sit to the artist without " making themselves up for the occasion; thus: a novel style of arranging the

A subtly worded swinging sign such as the one shown above is a dynamic example of the growing use of photography in advertising—and, incidentally, vice versa. If the sign can possibly be fixed so that as it swings it hits the inattentive passer-by in the back of the neck, the message impresses itself even more forcibly.

" hair, divesting the face of whiskers, beard or moustache, or making other
" changes (e.g., adding whiskers, beard, or moustache—ED.) will so palpably
" alter the general appearance of the sitter as to render recognition a task of some
" difficulty. . . ." With these instructions we are heartily in agreement, and
would indeed suggest that the admonition be suitably lettered, framed and placed
in a prominent position in the studio. Elaborate preparation such as is shown
in the photograph on page 14 is to be deplored.

When the writer, however, goes on to say " . . . All constrained attitudes and
" unmeaning expression of features should be also avoided. When accessories
" are introduced by way of accompaniment to the portrait, care should be taken
" that these are characteristic of the sitter's tastes and habits, and reasonable in
" themselves. Thus, placing a book in the hands of a person who is notoriously
" illiterate is an obvious solecism ; as is also representing a female striking a
" guitar, who does not know a note of music. . . ." With this dictum we positively *do not* agree ! A brilliant, if somewhat eccentric contemporary æsthete
has remarked that Nature imitates Art and, concurring as we do in this observation, we urge all photographers to stimulate their sitters to higher attainments
by giving them an appropriate vision of themselves to live up to. By all means
provoke the unmusical to " strum the lyre " and the illiterate to read books.

A final remark by this writer on photography serves to emphasise that he is not
altogether a *practical* man when he says " . . . When persons are having their
" portraits taken, it is a good plan to divert the mind by recurring to some agreeable
" incident in their past life, the thoughts of which will impart a pleasant and
" natural expression to the features." We can only remark that selective clair-
voyance is not yet a normal attribute, even of the experienced photographer.
Refractory subjects can always be clamped in position and left to cool off—this
diagram shows a model we have used for years with invariable success.
Incidentally, this same apparatus can also be used to maintain the
" Hand-stand " position during a long exposure ; the
head being placed on the upholstered seat and the right
thigh clamped into the top bracket.

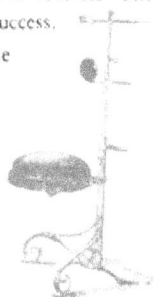

This is, no doubt, the sort of jiggery-pokery that the unknown critic quoted on page 13 has in mind. That such scenes are taking place every day in the dressing rooms of the three unscrupulous photographers shown here I do not doubt, but, fortunately for the prestige of the profession as a whole, such practices are definitely on the wane.

14

It took a long time to convince the young ladies shown in my composition " Five-Finger Exercise " of the educational value, both to themselves and others, of earnest cultural scenes such as this. Apparently they all thought the Harp was an illegal instrument— a fallacy doubtless induced by the colloquial phrase " Don't harp on it ! "—but when they saw the point they took it, as it were, to their bosoms and it was most difficult to get them back to ordinary bread-and-butter work.

After spending several jolly hours looking for freshwater crustacea a bare half-second at f8 was more than these land girls could stand. Repeated invocations to watch for the dickey were unheeded and it was only by draping my "Carlyle Cloak" with water-weed and advancing slowly on all fours that I was able to arouse their interest at all.

Outdoor Photography

It is not always appreciated—even by the experienced *Home* photographer—that OUTDOOR PHOTOGRAPHY, in common with other sports such as Aeronautics, Beagling and Cricket, has its own distinctive dress. Some people (e.g., the late Thomas Carlyle) have an innate flair for appropriate photographic dress. Others, and it is to these that my remarks are primarily directed, could obviously do with a little kindly guidance in the matter. Pausing but a moment to cast a disapproving eye at this example of what *not* to wear, we pass quickly, as is our wont, from adverse comment to constructive suggestion.

THE "CARLYLE" CLOAK.

It is safe to say that a voluminous black cloak of the type shown above is the garment *par excellence* for the outdoor enthusiast. Fitted with eight galvanised iron tent pegs and a detachable bottom-curtain, or brailing, the Carlyle Cloak

" Deerstalker " lens cap in Balmoral tweed.

" Gorblimey " lens cap in Lambeth tweed.

Lady enthusiasts wearing the new season's " Carlyle Cloak " with Junkers pattern Lens Cap to match. Only photographers with enough experience to know just what high temperatures can be reached when working at high pressure inside a closed cloak will fully appreciate the point of the abbreviated under-garments.

makes an admirable Portable Dark Room. To preserve privacy when working it is customary to run up a small red pennant embodying a suitable caution, such as :

"CLOAK ROOM FULL,"   or

"MAN AT WORK."

Inside this roomy enclosure it is possible to develop practically anything, and, if the weather should prove inclement, one can sit quietly inside with the flag up and remain philosophically isolated for hours. For summer wear, lady photographers sometimes affect an additional light "*over-cloak*" of flowered chintz or gaily striped organdie ; but this frivolous practice is usually deplored by the more serious male votaries of the art. Last year, indeed, we saw a particularly elegant model ; it was composed of cloth *of the new shade of pink*, soft and delicate, and was trimmed with bands of swansdown, looped over at regular intervals with black velvet. The trimming was carried down *one side* of the front, over the shoulder, and in a diagonal direction across the back, down to the bottom of the cloak. The mixture of the black, white and pink was very happily conceived. For a brunette, a scarlet cloak arranged in this manner would be equally stylish. It is not advisable to roam too far afield when *totally* inclosed in a cloak, as, in addition to the fact that it is impossible to see where one is going, the spectacle has an extraordinarily irritating effect on the average yokel or bull.

Before leaving the topic of appropriate wear a word or two about LENS CAPS is not out of place. Here, we are pleased to say, there is more latitude for individual taste and there is positively no optical reason why a nice "Deerstalker" pattern in Balmoral tweed (with characteristic side flaps and button at top) should not give as good a result as the plainer, peaked or "Gorblimey" variety. For the ladies what could be better than a crêpe bonnet, trimmed with three bows of graduated lengths falling down on each side, with a bunch of daisies and white lace at the top. A feather placed on each side and fastened at the back of the curtain, almost concealing the crown ; the curtain being made of silk, trimmed with lace, and the bandeau inside of white and coloured daisies ? A well-fitting Lens Cap or Bonnet is an absolute necessity for the hardy outdoor worker. In my opinion, the model featured in the picture below is both practical and becoming.

For strenuous outdoor work, of the hunting-shooting-fishing variety, what could be more efficiently attractive than the outfit shown here ? Wearing a concealed Hat-type camera and lens cap combined, with the tuck-in bloomer pattern Carlyle Cloak, this young lady is ready for instantaneous exposure, if need be.

STREET WORK.

Although this branch of photographic endeavour is cruelly limited by Mr.

Gladstone's " Street Nuisances and Performing Animals Act " of 1884, it is still
practised—even if somewhat furtively. And so, once again, I turn to the Hand
Camera Manual for moral support and technical weight. . . . " For ordinary
" landscape and marine work—as a rule, at all events—there is but little need for
" any concealment of purpose. But in the street it is desirable for two reasons
" that the camera should not be detected. Firstly, because of the attention it
" will attract, and secondly, on account of the set poses that will follow.

" Every endeavour, at all events, should be made to prevent the people in the
" scene knowing that they are ' going to be took,' or else they will all be found
" standing like plaster images staring at the camera for all they are worth. In any
" study of street life, character, or incident, natural grouping is essential. If it be,
" say, a fruit stall with customers, it would not be well rendered by each figure
" therein being represented as looking straight at the camera. It is not natural,
" it is not business.

" Upon this subject one word of advice. The beginner must not trust to any
" attempts at concealment in the design of the camera itself. The day is long
" past when even a plain black box or a bag will deceive the public.

" No, rapidity of action and secrecy of movement will effect the purpose in a
" more reliable fashion. The camera should not be raised or pointed until the
" exposure is possible, and this is where quickness of action comes to the front.
" If something intervenes to prevent the exposure, the camera should be dropped
" at once. Above all things, the worker should endeavour to forget that he has
" anything of the kind with him, because if he pays attention to the camera, other
" folks will do the same very quickly.

" There are many little wiles and tricks—in fact, the up-to-date hand camera
" man should be a deceiver of the deepest dye—such as lighting a pipe or cigar,
" buttoning a coat, taking off the hat to wipe the forehead, blowing the nose,
" looking into a shop window, etc., etc. Anything and everything in fact to cheat
" the public, to deceive them as to purpose. A friend to talk to is also occasionally
" useful, but nine times out of ten he gets in the way, and is better left at home
" to mind the baby. It is also a mistake and a very common one, to regard the

" scene or objects too long or too fixedly.   The worker should avoid being seen
" to possess an interest, though he may keep the matter under close observation.
" ' *Let not your left hand know what your right hand doeth.*' " (Verily—ED.) .

THE BEST TYPE OF CAMERA FOR OUTDOOR WORK.

Naturally, in addition to the foregoing information, a certain amount of consideration should be given to the choice of a camera best suited to one's individual needs.

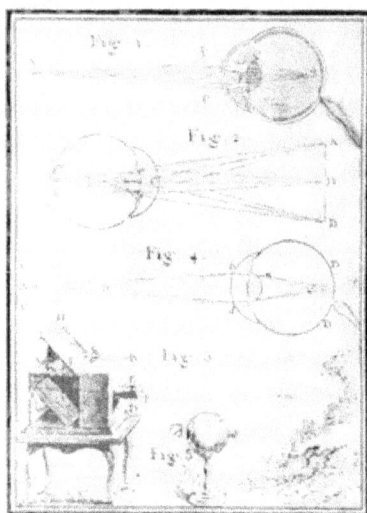

The Box Type Reflex Camera (Fig. 5), shown above has some obvious disadvant-
ages and despite the maker's attempt to distract attention from its structural faults
with an adjacent drawing of conjugal foci one should not be blinded to the physical
difficulties of transporting such an apparatus " o'er hill and dale " . . . or even
just " o'er hill."   The smaller Spherical Model (Fig. 3) is more easily portable
but suffers from the disadvantage that when *in situ* it has to be constantly guarded
from small boys who, to use their own playful expression, " want to kick the ball

off the stick " . . . shortsighted golfers have been known to take a swipe at it with a driver.

The writer favours the type of ALL-PURPOSE Camera shown here.

Fig. 1.

Fig. 2.

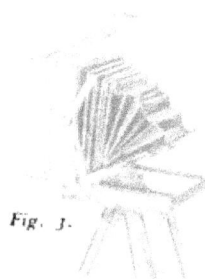

Fig. 3.

Fig. 4.

As can be seen, it can be used for Hand (Fig. 4), or Stand Work (Fig. 3), Hand-Stand Work (Figs. 2 and 4 combined and reversed), Short Focus (side angle) Lens (Fig. 1), Long Focus (Fig. 2), and No Focus (Fig-leaf) ; it can also easily be adapted to carry four medium sized sandwiches and a large apple and is much less conspicuous than the LUNCH-BOX Camera shown below.

An early Lunch-box Camera— now superseded by the type shown in Figs. 1, 2, 3 and 4.

SPECIAL PATTERNS.

We quote from the " Hand-Camera Manual " :—" The reader should understand
" that many of the wonderful inventions we read of in the non-photographic
" press are rarely, if ever, to be found on sale. Such a one, for instance, as the
" Soda Water Bottle, which snap-shotted a man in the act of drinking.

" These inventions are creditable—to the journalist. But there are some really
" pushed as downright useful things, which are the merest toys in reality. In the
" majority, to start with, the pictures are too small to be of any value. I have no
" wish to offend, but I have certainly been surprised at the absolute rubbish
" offered to the public, not only by the outsider but by the photographic dealer.
" The former I can understand, for he may not even know that it is *desirable*
" that the camera should be light tight. His business is to sell the cameras. But
" the photographic manufacturer or dealer, really must know sometimes that he
" is putting forward to photographers mere toys. At the same time there are
" novelties which are capable of first-class work, and with two of these I propose
" to deal.

" THE HAT CAMERA.

" There have been so many humorous attempts at concealment of a camera that
" to mention a ' hat ' camera would, at first sight imply something similar. But
" this is not so, because a full size quarter-plate is used, and covered as well. The
" apparatus is simple enough notwithstanding, and can be fitted to any stiff form
" of hat—the round billycock, or the chimney pot of the Metropolis. Even the

Fig. 1.

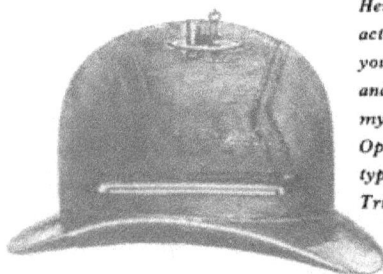

*Here we have a model (in scale, in glass) of the
actual Felt-Hat Camera used by the author in his
younger days to take the series of Instantaneous
and Moving Object photographs used to illustrate
my article on that subject (see page 30). The
Operatic-Hat Camera is a variation of the above
type and is used, in conjunction with the Umbrella
Tripod, for general espionage and Night Club work.*

" small hole necessary to permit the clear view of the lens can be so neatly cut, and
" the part fitted so in as to attract no attention. The camera consists of a bellows
" body, which lies flat in the ordinary way, but is extended by wires when required
" for use. Special firm dark slides, of course, are necessary, and in addition there
" is a focusing screen. The shutter is quickly got ready by placing, by a half-turn,
" the release spring into its receptacle in the front of the hat. The camera alone
" which is outlined in Fig. 1, weighs by itself 2½ ozs. only. The method of use
" is shown in Fig. 2.

Here we have a model (inaccurate, in wax), of the author with his Felt-Hat Camera deceptively poised in an attitude of old-world courtesy, whilst the index finger of the right hand hovers expectantly over the button. Further to conceal his intentions his left eye is seen to be focused on a point due west of where the camera is pointing— and presumably his right eye is focused on a point due east of same. It is doubtful whether the subject had the slightest idea she was being photographed.

" THE ' BINOCULAR '

" This is in the shape of a field-glass, one lens being used as a finder, and the other,
" of course, for the exposures. When charged with plates it only weighs 19 ozs.,
" and is arranged for twelve exposures upon plates measuring 2½ × 1½ inches.

Fig. 1

Fig. 2

" Some idea of the arrangement and changing method may be gleaned from the
" above (Fig. 2) illustration. It is very neat, effective in use, and the results
" shown are good."

The Patent "Hotel" or Keyhole Camera.

This remarkable invention which has done so much to facilitate the gathering
of *in camera* evidence first came into the news of the world when the late Lord
Justice Bunkum naively enquired " . . . Why were the defendants enclosed in
" a bottle at the time ? " (laughter in court). The prosecuting counsel then
explained that the unusual shape was caused by the *keyhole* through which the
series of photographs had been taken. Before showing a selection of such photographs—extracted from the series that resulted in a conviction in the famous
Entwhistle *v.* Arsenal case—I propose to give a brief description of the ingenious
apparatus of which I happen to be the modest inventor.

A cardboard notice which reads " QUIET, PLEASE " . . . " MEN AT WORK " . . .
" OUT OF ORDER " . . . or, indeed, anything else you like, is securely attached
to the back of an ordinary Box-Type Camera. The camera is then loaded in
the normal way and attached to the door by means of two or three of my patent

GRIPWELL SUCTION PADS (4 6 *per box of twelve assorted sizes, obtainable from any reputable ironmonger or chemist*), so that the LENS APERTURE coincided with the keyhole. Next, the doorknob is quietly unscrewed and replaced with a gilded india-rubber bulb (for working the camera shutter, presumably—ED.). Finally, and unless an existing hole made by the hotel staff for private reconnaissance can be located and utilised, a VIEWING HOLE must be bored through the door. This need not be very obtrusive—¼ inch diameter is usually quite sufficient—and we suggest that the room number can often be used to provide local camouflage; e.g., if the number contains 6, 8, 9 or 0 the hole should be bored thus

whilst if it does not contain any of the above figures the best position in which it can be determined by studying a Moorhen in its natural surroundings, or a Purple Patched Wood Louse hiding in a stamp album

When fixed as described, the Keyhole Camera is ready for use, but immediate use is advisable since too much loitering about one's door may look suspicious. Simply peep through the Viewing Hole, and when ready, simply press what everyone else thinks is the door knob. The resulting Key-hole Studies present many interesting features, some of which appear in different sections of this book. Unfortunately, many features it is impossible to touch on anywhere else, but which, in the original series, were missing from the collection when it reached the mounters.

Moving Object

The first requisites for moving objects are a green baize apron and a wide vocabulary : from then in it is a question of *you* versus the object. This series happen to be an example of prolonged indoor work, but the same general principles apply outdoors, although the exposure will probably have to to be cut down to conform to police regulations.

A contemporary authority, who would deem to be interested in rather faster subjects than I am, has the following comments to make : " . . . That wretched phrase " ' *Instantaneous* ' gets in its fell work every-" where, and I must warn the beginner " against a very common failure, that of " firing at objects which are in reality moving " at too great a speed.

Photography

" The following table will show at once
" what can be done and what should be left
" undone :—

" Man walking   5 miles per hour.
" Vessel travelling at   20 knots per hour.
" Finish of Cycle Race   30 miles per hour.
" Express Train   50 miles per hour.

" To find the distance the object will move
" upon the plate is it only necessary to multi-
" ply the focus of the lens in inches by the
" distance moved by the object in the second,
" then divide the result by distance of the
" object (from the lens) in inches, and
" finally divide by the speed of the shutter.
" For example, I will take the finish of a
" cycle race under ordinary camera condi-
" tions. The lens of 5½-inch focus, the
" shutter working at the 1 30th of a second,

' and the object 10 feet away, the calculation would come out

$$" \quad 5\tfrac{1}{2} \quad 44 \quad 242 \quad 120 \quad 2 \text{ inches per second}$$

" Now as the shutter works at the 1 30th of a second, the movement upon the plate
" would be a fraction over 3 30th or 1 16th of an inch. The resulting photograph
" would be a curiosity.

" In the above example the conditions are those of the majority of hand-cameras,
" as although the shutters are often put down as working at the 1 100th of a second,
" or at even greater speed, considerable discount must be allowed upon these
" statements. The limit of movement upon the plate, if anything like a sharp
" image is desired, is the 1 100th of an inch. So that by working backwards we
" can find the shutter speed required or the distance from the object actually

A glaring example of " Footlight Exposure "

" necessary. The speed would have to be increased to the 1 200th of a second,
" or with the same shutter (working at 1 30th) the camera would have to be
" 2,106 feet 10 inches from the object, which is too far to walk.

" I trust these tables will prevent the beginner from wasting plates in the absurd
" fashion that I have seen done on many occasions. Of course the movement is
" calculated full broadside on, an object coming towards or receding from the
" camera is a much easier task."

Exposure

Brushing aside the rather prudish attitude that condemns exposure of any kind
we would say that the exposure should always be regulated to suit the subject—
and, of course, the lighter the subject the greater should be the exposure.

It is my experience that the subject has to be coaxed into the appropriate degree
of exposure ; and, as a general rule, it is desirable that the lighting should be
regulated so as to give the maximum encouragement.

Some subjects expose better with full lighting (e.g., limelight or footlights), and
others prefer subdued lighting—preferably pink : whilst difficult cases refuse to
expose anything worth looking at unless the light is out. In such cases *infra red*
is not *infra dig*.

When gazing at a well exposed subject do not breathe too heavily on the view-
finder as the resultant haze tends to rob one of that *savoir faire* which is so essential
an ingredient of indoor photography.

TIME EXPOSURE.

Once more according to Mr. Welford . . . : " It is rather awkward to refer to
" time exposures, that is, exposures needing a support for their accomplishment,
" because it is really not hand-camera work at all. But there are occasions when,
" for want of light caused either by the dullness or lateness of the day, or by the
" scene or object itself, prolonged exposures are necessary.

" With practice a full second is easily managed, especially if the body be utilised
" to the best advantage. One great point in this is to first steady the body, by

In this example of landscape with heavy foliage (half a second at f8) we see a pair of conscientious students looking for a fixed support suitable for a really long exposure. As no fence, gate, wall or rough erection of stone or wood was available they just had to make do with a hedge.

" sitting down, or leaning against a support. Holding the breath during the
" exposure is recommended by some, but I have not found it of much assistance,
" as the strain of so doing is as bad as the breathing.

" For longer exposures I should strongly recommend a fixed support, and this is
" often obtainable by search (see illustration on page 34). The top of a fence,
" gate, or wall, rough erection of stone or wood, for instance.

" Tripods may, of course, be pressed into service, and indeed, there are several
" varieties upon the market made specially portable for this very purpose . . .
" In buildings, chuches, etc., there are many opportunities afforded of local
" support, such as pews, two chairs placed back to back, etc."

With such a wealth of practical suggestions I can only agree.

OVER EXPOSURE.

With some subjects it is practically impossible to prevent over-exposure : I have
found that complete immersion in lukewarm hypo until the bubbles stop rising
is as good a cure as any. Over-exposure is, on the whole, more desirable than
under-exposure : unwanted detail can usually be eliminated by deft retouching.
We show, on page 39, a few specimens of rather obvious over-exposure.

PATCHY, OR PARTIAL EXPOSURE.

Where uninteresting portions of the subject appear to be over exposed, whilst
other more decorative zones are obscured, it is often a good idea to ask the sitter
to " watch for the dickey." However, it is not wise to keep encouraging the
subject to " watch for the Dickey " as, if it fails to appear, disappointment and
inertia often result. In obstinate cases a teaspoonful of Butyl Chloride in a cup
of steaming hot cocoa will usually do the trick.

UNDER EXPOSURE.

This is a common fault with beginners and it is up to the photographer to guide
the subject. Don't rush matters—a few well-chosen formulæ and a tumblerful
of neat alcohol are usually all that is required. If the light is too strong, draw
the blinds : if the light is not strong enough, draw the blinds and light the
lamp—remember, incandescent exposure of any kind is frowned on by the
authorities. It is advisable to remove the lens cap before letting the hair down.

DURATION OF EXPOSURES.

One of the easiest questions to ask and the most difficult to answer, is " what exposure is suitable ? " The whole matter is so governed by various factors that it is next to impossible to give any direct answer.

The factors to be considered are the following :—

(1) The nature of the subject.   (4) Speed.
(2) Strength of the light.   (5) Development.
(3) Aperture.

The great difficulty is to bring home to beginners the considerable effect of variation in any one of these points. As a rule they do not grasp the importance of, say, (1), (2), and (3). Perhaps the following tables by Prof. Burton and Dr. Scott will be of assistance in the matter :—

COMPARATIVE EXPOSURES—(BURTON).

	f8	f11	f16
Sea and sky	1 40	1 20	1/16
Open landscape	1 12	1 6	1/3
Landscape with heavy foliage ..	½	1	2

It will be observed that each stop as it decreases the aperture doubles the exposure. Thus, an exposure of one second with f8 would be two seconds with f11, and four seconds with f16. These are given for ordinary slow plates, and should be decreased by one half at least for the more rapid brands.

Thus, for a well-lighted landscape 1 24th of a second only (with an aperture of f8) would be required. The table is calculated for bright lighting.

Another factor is the strength of the light. This is a most unreliable one to judge until experience comes to the rescue. The difference in the actinic power of the light, even in bright sunlight, between the morning and afternoon is great.

The photographs (page 38) give some idea of the powerful dramatic effects that can be obtained by cleverly varying the exposure. The ivy leaves were put in freehand when it was all over and had it not been for the cat in the tin hat taking up so much room this caption could have been in its proper place under the picture—instead of here.

Hour of Day. a.m.	p.m.	June.	May, July.	April, August.	March, Sept.	Feb., Oct.	Jan., Nov.	Dec.
12 noon.		1	1	1½	1½	2	3½	4
11	1	1	1	1½	1½	2½	4	5
10	2	1	1	1½	1½	3	5	6
9	3	1	1¼	1½	2	4	12	16
8	4	1½	1½	2	3	10		
7	5	2	2½	3	6			
6	6	2½	3	6				
5	7	5	6					
4	8	12						

It will be noticed that in any month the best time is between ten and two, when the light is strongest and with least variation.

All the evening exposures present another difficulty, as with a yellow sunset the necessary time would have to be increased. Possibly to a beginner the following table will be of first utility, the plate being of the rapid variety and the light good.

	f8
Sea and sky	1 100
Street scenes (open)	1 50
Landscape (open)	1 20
Landscape (heavy foliage) ..	½
Interiors	*Anything from 3 or 4 minutes to several hours, according to the amount of light.*
Outdoor portraiture	*Same as open landscape.*

If this last table be worked in conjunction with the others, it will be fairly simple to make comparative exposures, taking the light as :—

Bright sunshine	1
Cloudy bright	2
Dull	3
Gloomy	4

As a rough guide, if the lens in the hand-camera has an aperture of f8, on a bright, sunshiny day, with a rapid plate—a street scene about midday in June will require 1 50th of a second. The beginner can make all other calculations from this, as a basis. But his own results will tell him more in this direction than I can.

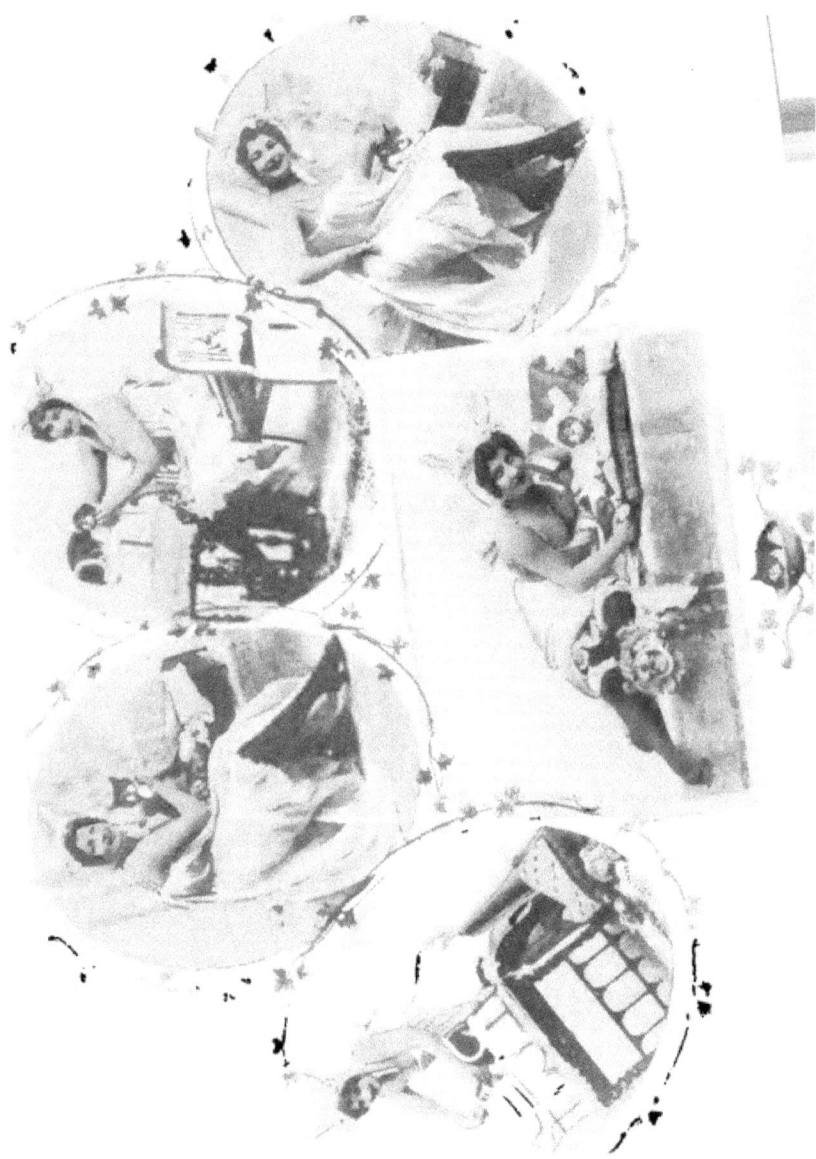

Development

This is rather a personal matter : every photographer is entitled to his own ideas, and, generally speaking, it is only after a long experience of trial and error that one learns to tell at a glance when the subject is properly developed for the particular purpose in mind.

It is best to get this rather tedious business over during adolescence if possible, thus leaving the adult years free for selective rather than promiscuous experimental work. Fashion, of course, plays a paramount part, and a careful student of developments cannot fail to observe that every type has its period of popularity.

It is customary to unload in the dark-room ; and one of the first things to remember is that, however safe the dark-room light appears to be, there is nothing to be gained by premature exposure. It is sound practice to work in the dark as long as possible : always be ⬛⬛⬛ wary of red lights.

Here are some fine examples of the good results that can be obtained by a discreet application of my " Land Development and Reclamation Scheme." There is not the slightest doubt in my mind that proper exposure under sunny conditions can do a great deal to eradicate troubles that are often quite wrongly attributed to faults in development.

DEVELOPERS.

The older types of developers such as Indian Clubs, Night Clubs, Dumb-bells and the Sandow Course are gradually being superseded. New problems call for new techniques and I am very pleased to be able to report that several much less strenuous solutions have already been found.

THE SOLUTION.

It will be better for the beginner to buy his solution ready prepared. There are various safe proprietary brands on the market, and the individual is well advised to find one that suits him and to persevere with it. After mixing the solution according to the directions on the packet the subject should be laid face upwards in the bath and gently rocked backwards and forwards . . . backwards and forwards . . . backwards and forwards.

FIXING AND WASHING.

When removed from the solution, a white or milky appearance will be more or less visible—it is customary to describe this as the unacted-upon part of the silver bromide. Anyway, whatever it is, it is the purpose of the fixing solution to eliminate it. First wash thoroughly in water and then dip in hyposulphate of soda (hypo for short). Up to this stage all the work has been done either in the dark or by the correct yellow or ruby light, but after fixing it is quite in order to pull the blinds up, although washing should continue for quite a long time. Beware of fog . . . too much soda in the bath or over-exposure are common causes.

OVER-DEVELOPMENT.

Very little can be done about really exuberant over-development, although, to some extent, its worst effects can be modified by local reduction. The choice of general reducing agents is varied ; Lord Byron (early photographer—Ed.) is said to have favoured saturation with carbonic acid and water, whilst at least one well known proprietary brand would seem primarily to be composed of a discreet mixture of Magnesium and Sodium Sulphate. Constant friction with a rubber roller or squeegee in the region of the affected part or parts has been known to give quite good results and the heat generated is sometimes sufficient to boil a kettle . . . if you want to boil a kettle. As a general caution we would advise enthusiasts against using any brand of local reducing agent that is known to encourage worms as a by-product.

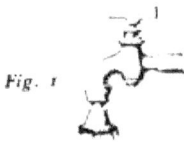

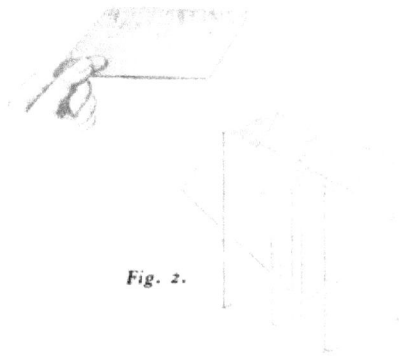

Here you see alternative washing devices (1) The Rose Sprinkler, (2) The Washing Trough, (3) The Steps, Pipe and Barrel method. Of the three we definitely prefer the latter, since the same apparatus, can, in lieu of the Indian Rope Trick, be used to get the subject into the appropriate condition (position ?—Ed.) for Megascopic exposures (see illustration and details on page 61).

UNDER-DEVELOPMENT.

We will avoid entering the controversy regarding what constitutes *proper* development—nowadays the standard on such matters would seem to be an arbitrary one, fixed from year to year by R.A.'s and corset manufacturers : sufficient to say that it was not always thus ! P. P. Rubens, a well known, if rather Flemish, photographer who flourished round about the reign of Charles I, was a staunch supporter of Over-Development in subjects of all sexes. (*See Pictures " Toilet of Venus," and " Hercules "*). Whilst his contemporary Theolocopuli Domenico (surnamed El Greco) would seem to have favoured Under-Development to the point of skinnyness. As far as is known, the connoisseurs of the time—who are briefly referred to as " rakes and libertines " in school history books—shared this eclecticism.

It is my considered opinion that, as an exponent of Over Development, P. P. Rubens rather overdid it. Too much of this sort of thing is guaranteed to make even the toughest æsthetes boggle a bit and photographers are urged to pursue the happy medium—if they can find one.

One thing is certain—even experienced amateurs have difficulty in defining precisely, (1) what *causes* Under-Development ; (2) what *constitutes* Under-Development. Of the latter I can only say that I regard the whole thing as a matter of taste and when in doubt I use drapery, or soft focus, or both. The time may come when questions of this kind will be decided for us by an authoritative body of scholarly experts sitting in dignified solitude in some remote city, such as Los Angeles . . . until that happy day it is *chacun a son goût*, as they say over the water.

I don't want to bore you, dear reader, but here is yet another example of what is being done all over the country under my "Land Development and Reclamation Scheme." I think you will agree that, despite slight over exposure and temporary lassitude (induced, no doubt, by too close an application to the fascinating sport of making hay while the sun shines), very good work either has been or is about to be, done here.

As to what *causes* Under-Development, the answer is, simply . . . not enough developing; and the best remedy is, of course, more development. I always warn pupils of mine against taking specious promises of the " You, too, can have a body like mine " genre too literally; in my opinion vigorous physical exercise tends to produce knobs and bumps rather than flowing curves. It is my contention that the finest form of all round physical culture takes the form shown in the illustration on page 8, in which four old students of mine are initiating a new recruit. As you can see for yourself, the beginner is the rather meagre miss with her back to the camera; the others have obviously been at it for years.

If, in spite of all this, outdoor exercise is still preferred, a leisurely and decolleté course of static-boating as practised by my second-year students on page 45 often has excellent results; although the rapt expression on the face of the oars-woman rowing rhythmically in half an inch of water with an oar *sans* blade is a sad indictment of the mental condition to which persistent exercises of this kind inevitably reduce one.

The Dark Room

It is a good idea to arrange for the Dark Room to be situated in a fairly inaccessible corner of the house: complete privacy is essential. The " Carlyle Cloak " mentioned in the chapter on Outdoor Photography can be adapted for indoor use, in conjunction with a tea chest and a kitchen table, as shown in Fig. 1. All bottles should be clearly labelled Poison and, unless discreet collection can be arranged, when empty they should be broken up into small pieces and dropped down the sink.

The Drying Cupboard should be large and roomy with plenty of space for plates, bottles, light snacks and, if possible, a small settee of the portable-collapsible type. Experienced photographers usually regard the Drying cupboard as a second bastion of defence and equip it to withstand at least a month's siege.

There is no need to go into long details about Developing Trays—the writer can only say that he personally, prefers ones with white transparent bottoms, since their cleanliness can be more readily ascertained. Apart from usefulness in disposing of the empties a sink is regarded as practically a *sine qua non* for washing

One cannot avoid remarking that transparent spots (side section on "Common Defects of Negatives") are again in evidence in the picture. "Static floating"; caused no doubt by the large clots of chloride of sodium which can be seen hovering about in the sky in a predatory sort of way. Notice, too, the flowered organdie lens cap and summer-weight "Carlyle Cloak" worn by the young lady on the left. A certain amount of over-exposure was unavoidable, although the shot was taken in brilliant sunshine with an exposure of only one-hundredth of a second and an aperture of f8.

Only by biting away the corner of the apparatus (and the lobe of his right ear—Ed.) was I able to get this interesting view of a fellow enthusiast at work inside a domestic adaptation of the " Carlyle Cloak."

unless the method suggested by Wratton and Wainewright for using alcohol be employed. I have no sink. It has been found that a red or amber light is conducive to the best results, and I thoroughly agree with Captain Abney, R.E., F.R.S., etc., who, on page 14 of his book " *Practical Working of the Gelatine Emulsion Process*," says : " . . . For our own part we prefer light to come from " about the height of one's waist, since all operations can then be distinctly " seen. . . ."

Another writer, Walter Welford, has some useful general hints in his little book " The Hand Camera Manual "; on page 87 he says : " . . . If the room be a " small one it will be much better to have artificial light outside, as the close " atmosphere of a small room is certainly not conducive to health. Roughly " speaking the other requirements of a dark room are, a table, a receptacle for " water and a jug. But if a sink be available so much the better, as the inconveni- " ence of a pail will soon be discovered. Water direct from the tap is also a great " convenience, as it is frequently required. A shelf for bottles, etc., should " also be provided. . . ."

This is, of course, a Developing Tray or Bath : for those who like to splash about or play with cellu-loid ducks the larger model shown on page 47 is the only logical answer.

Toning, Fixing and Washing

The details of the procedure must be determined, to a large extent, by the individual photographer's preferences in the matter of toning bath and paper. Chloride of gold, in conjunction with other chemicals, is the most generally used toning agent, giving warm-black, red-brown and red tones ; whilst Platinum and Uranium are sometimes used to give sepia tones. Chloride of Lime should

A capacious bath is absolutely indispensable for photographic work. Practically everything needs washing, fixing, or soaking at some time or another and in an up-to-date bathroom like this all these processes become a pleasure. The fact that I happen not to like the wall-paper or the tattooing on the side of the main bath is mere æsthetic whimsey and certainly does not blind me to the many excellent and practical features of the plumbing : in any case, much of the processing is done either in the dark or by red lamp light.

not be used as a substitute for Chloride of Gold, as, in addition to being useless for the purpose, it causes pimples and gradually dissolves the fingers. In my opinion the best paper for the beginner is the Gelatino-Chloride variety, on account of its easy manipulation and the range and tone of finish obtainable. The other papers can easily be tackled later on. The amateur could not do better than be guided by the detailed directions included with each packet of paper. According to the Hand Camera Manual : " . . . There are two main principles " adopted in which the operations differ. These are termed the COMBINED " BATH and the SEPARATE BATHS, and briefly, the operations may be thus " described :—

COMBINED BATH.	SEPARATE BATHS.
Subject immersed without washing.	*Subject washed.*
Toned and Fixed at the same operation.	*Toned.*
	Rinsed.
Washed after Fixing.	*Fixed.*
	Washed after Fixing.

" The Combined Bath is certainly less work, and it is adopted by many. It is, " however, a little more tricky in its nature and is not so certain in result in a " beginner's hands." As a matter of interest, I would like to point out that the model shown in the illustration on page 47 is definitely *not* suitable for Combined Bathing ; and I don't think a great deal of the paper, either, having an old-fashioned preference for the grapes-crawling-up-a-trellis pattern. The Hand Camera Manual goes on to say that the subject is Toned (or Toned and Fixed) by immersion in the solution in the tray, which is kept in constant motion. They . . . " must not stick together in the tray but be constantly changing " position by means of the fingers. They then receive a slight rinse in water " and go into the Fixing Bath, where they remain for about 15 to 20 minutes. " After thorough washing they are passed through a bath of Alum to harden " the film, and lastly dried."

MOUNTING AND FINISHING.

For ordinary mounting the subjects are taken out of the water and placed on a linen sheet, another sheet is placed on top and the hand rubbed firmly over the

Although I have consistently advised the use of large and roomy baths—particularly for Combined Bath work, as detailed on page 48—I cannot help remarking that this young enthusiast has gone a bit too far. Fishing about in roods and fathoms of water on the off-chance of finding a couple of half-plates is not my idea of photographic efficiency; and why the young lady should look so pleased with herself for having found what looks suspiciously like a prawn, is really beyond my ken. A couple of hours at f16 is what she needs!

top sheet ; this removes surplus moisture. They are then gathered together into a neat pile and laid face downwards on a sheet of clean paper. Then the Mounting Medium is applied and the print is rubbed down on to the mount with the hand, over a piece of blotting paper. If it is desired to *dry* mount they must not be placed between blotting paper, but laid on a piece of glass, cloth or paper, and left uncovered until dry. For the special surfaces a *Squeegee* is required. This consists of an indiarubber roller mounted with a wooden handle. A piece of vulcanite, enamelled iron (ferrotype) plate, or plate glass for the highly glazed surface ; and a fine ground glass or matt surface celluloid film for the matt or dead surface is also necessary. These must be carefully cleaned in warm water, polished with a soft silk handkerchief or wash-leather, and when dry dusted over. with French Chalk (or Fuller's Earth—? ED.). When this is again dusted off, the subject is placed film down whilst wet upon either surface, a piece of blotting paper placed over it and the Squeegee applied vigorously. If left in a warm dry place they will strip off in a few hours.

How to tell Positives from Negatives

Generally speaking, negatives are darker than positives. But the whole subject is fraught with difficulty ; it is safer to say that most negatives are a prelude to

Perhaps the most important difference between a negative and a positive is that a negative is denser in the parts where a positive isn't. Here we see the author using a negative to print down mural on the walls of the Chapel-of-Ease at Stratford-le-Bow.

a positive, indeed, persistence will usually turn the most obdurate negative into a positive, whilst some negatives seem automatically to turn positive during the final stages.

When asked to give an opinion it is well to avoid being too definite . . . . why why give others the benefit of your hard earned experience. The picture on page 59 is a case in point ; most critics said that the subject was obviously a negative—whereas I was able to affirm from my own experience that she was emphatically a positive. Some people spend a lifetime producing nothing but negatives—this shows a deplorable lack of versatility.

It is sometimes possible to combine negatives and positives in the same picture. The Beach Scene (this page), is a case in point ; here we have three positives (back row) one negative (front, left), and an indecisive (front, right).

This example of combined positive-and-negative print was taken under rather trying circumstances and I regret to have to admit that the old school colours on my boating hat were observed by one of the subjects (rear, left) before the full exposure had been completed.

It is perhaps significant that Louis Jacques Mande Daguerre, the first successful French exponent of Photography developed a positive process whilst his English contemporary, William Henry Fox Talbot, was concentrating on negatives. Never was the Gallic temperament better apostrophised.

Common Defects in Negatives

BLISTERS—If blisters make their appearance it is probable, if the substratum be of albumen, that the solution is not sufficiently dilute. With some hinds of india-rubber blisters always appear. *The practice of tacking prints on the wall with a coal hammer is another prolific cause of blisters.*

TRANSPARENT MARKINGS—may be caused by handling the subject with warm fingers before immersion in water previous to development. *Handling with cold fingers has its own problems.*

A TRANSPARENT EDGE—will be caused by allowing the whole length of the edge of the subject to rest on blotting paper when drying in the drying-box. *The only consolation is that some subjects look better with transparent edges.*

A LACK OF DENSITY—is caused by the collodion being too thin, requiring more pyroxyline ; by an insufficient quantity of iodide ; by insufficient sensitizing in the bath ; or by too weak an alkaline developer. *Keeping the subject at school until Matric has been passed can only be regarded as a secondary cause.*

LINES—may be caused by a stoppage in the wave of developing solution, by removing the subject in the drying-box previous to complete dessication, or by an uneven flow of preservative over the film. *It is therefore a fallacy to assume that old-age and late nights are the only causes of this prevalent phenomenon.*

BLACK SPOTS—on the film may be due to the india-rubber substratum, and to dust on the plate. *They are sometimes due to indigestion, in which case they do not remain stationary, but move slowly in an oblique direction.*

TRANSPARENT SPOTS—may be met with when photographing near the sea. (*See lace insertions in bathing costume of subject standing by portable dark room, on opp. page*). They are probably due to the chloride of sodium which is held in

Practically all the defects listed in the accompanying article are apparent in this photograph of models resting in and around my special " Beach Pattern " portable dark room —but I still like it. Which only goes to show that great Art will always out, in spite of or because of, common defects.

suspension in the air. They rarely occur if the subject has been thoroughly dried finally by artificial heat a short time before exposure. *Many students regard this drying-out process as one of the best things about seaside photography.*

PINHOLES—may be caused by the solution of silver added to the developer dissolving out iodide from the film. If the preservative be not well filtered such defect may likewise occur. *Blast pinholes!* If the preservative used for the dry plate contains any substance only slightly soluble in the former, but more readily in the latter, then the latter should be flowed over the subject and allowed thoroughly to permeate the surface. A good washing under the tap afterwards is then necessary. If the preservative contains nothing soluble by alcohol, water should be applied in the first instance. *Quite a lot of defects can be traced to the too exclusive application of alcohol, regardless of solubility.*

Whether spirits of wine or water be the agent used for softening the film, great care should be taken that there is no stoppage in the flow, otherwise markings in the negative may become apparent. (A dipping bath or a flat dish is useful when water is to be applied.) The preservative must in all cases be eliminated from the film as far as possible before development commences.

Trick Photography and Montage

This is a much abused science. Genuine experimental work should not be confused with the spurious *carte postale* school which debases ingenuity by purely objective repetition. Most of my own researches into this fascinating branch of the photographic art have been essentially subjective ; indeed, practically all my original discoveries have been the direct result of persistent attempts to translate personal whims and fancies into photographic realities.

Such an attitude is necessarily both a limitation and an inspiration, For instance, although my Aunt Letitia (mentioned vaguely in another connection in this book) has a face that in general mass has a striking resemblance to a Jersey cow, close scrutiny reveals that she hasn't got quite as much hair in quite the same places as the head of that noble and productive animal usually has.

A realisation of this fact—bordering, I might say, on morbid fascination—prompted me to experiment and out of experiment was born this composite photograph in which, you will readily observe, the little differences between the two have been eliminated.

" Giving Nature a Helping Hand " is one of the most fascinating functions of photo-montage work. Aunt Letitia, who had a predilection for wearing odd blue stockings and rapping my knuckles when young, was the unwitting inspiration for one of my finest efforts in this direction. For those who think this kind of thing is easy I have only one answer—you're quite right, it is !

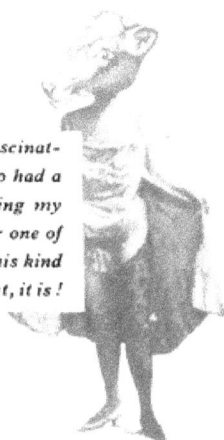

To call the photograph " *realistic* " would be wrong, since it undoubtedly flatters Auntie ; but it is my sincere belief (a belief supported by the opinions of many disinterested observers) that the visual *impact* of my photographic reconstruction closely approximates the effect of my Aunt Letitia *en personne* on persons (pun—? ED.). Again I would stress that scientific curiosity and not a desire for mere realism was the prime factor in all my experiments. On yet another occasion I remember being goaded into transposing a portrait (head and no shoulders) of my cousin Joe from its legitimate, if rather uninspiring, position in the family group reproduced on page 51 on to the torso of the " Idle Apprentice " in one of his moist Hogarthian moments. This experiment was the occasion of considerable resentment—Cousin Joe having practically no scientific curiosity— and the negative was unfortunately broken : the reproduction on the next page was made from the print that caused all the trouble.

To transport my cousin Joe into the midst of the gay little
scene above—so redolent of happy holidays at seaside boarding
houses—was a technical achievement of no mean order. I have
not the slightest doubt that photo-montage will eventually oust
all the cruder forms of blackmail.

The technical procedure adopted in the two cases already cited is now too well known to need elucidation, but the next example is rather more complicated.

FAMILY PORTRAIT—In addition to being an outstanding example of classical composition (based on the famous picture " Mountain Goats at Herne Bay," by Edwin Landseer) this photograph is also interesting as an experiment in remote control. The camera shutter was operated by an arrangement of wires, mirrors and gum arabic, a procedure of which Cousin Joe (rear, centre) strongly disapproved.

Briefly, the problem was this : how to concoct suitable photographic evidence for an old school friend who was seeking a divorce. The whole experiment was

58

rather delicate since the two parties most concerned—i.e., his wife and the intended co-respondent—had never been seen in each other's company ; and, indeed, had not, as far as was known, ever met.

On the face of it this set-up would seem to present insuperable difficulties, but after studying the problem from all angles I evolved a plan which, with all due modesty, appears in retrospect to have had the unmistakable hallmark of photographic genius. The stark simplicity of it was perhaps its most outstanding merit. I disguised myself as an itinerant exchanger of aspidestras for old trousers and armed with an amazing specimen of that domestic favourite and a convincing line of sales talk I called at Mr. X's bachelor apartment : as was to be expected, he came to the door in his trousers, upon which I complimented him heartily—meanwhile concealing the aspidestra under a voluminous black cloak.

Struck, no doubt, by my enthusiastic admiration of his nether garment he shyly invited me to tea ; upon which I threw open my cloak, revealing both my aspidestra and the fact that *I* had *no* trousers on. And then, with what I have been told is my most engaging smile, I offered him my aspidestra in exchange for his trousers : shivering to emphasise my necessity. Diffident at first, he gradually warmed to the idea and when I showed him what a touch of furniture polish did to the leaves he finally succumbed and took his trousers off. This was the moment I had been waiting for, and pressing the bulb of my camera (which, I forgot to mention, I had concealed in a large orange I was sucking), I secured a perfect photograph of Mr. X *in delicto aspidestrum*, and walked quietly away.

I hesitate to bore the reader with even more technical details of how an appropriate picture of Mrs. Y was secured (*for details see chapter on* " *The Keyhole Camera* ")—suffice it to say that, ultimately, the case was successfully concluded *in camera*. As a matter of interest I am pleased to be able to report that the ex-Mrs. Y. was so impressed with my series of composite photographs, a mild example of which is shown on page 59, that she made exhaustive independent enquiries which soon blossomed into true love, and they married and lived ever after.

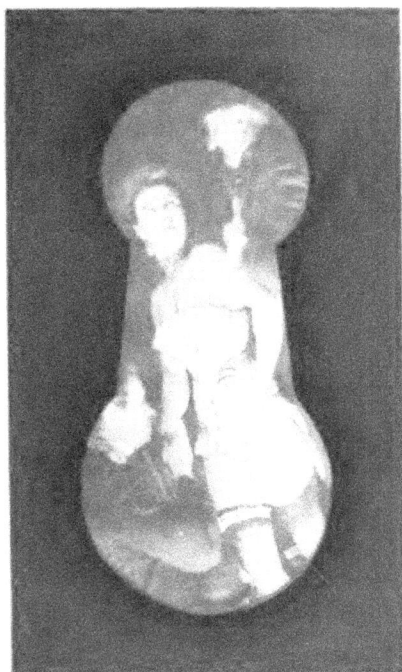

This is the composite " Keyhole " photograph referred to in my brief technical summary of the ' X and Y ' case. You will notice the slight obliquity—introduced to give a realistic air of anything-can-happen-now.

Miscellaneous Trickery

Some photographers, not content with exercising straightforward personal ingenuity, make a practice of employing all sorts of dubious technical gadgets, such as : Rumford's Photometer, Theodolites, Anamorphosis, Megascopes, Algebra, The Law of Diminishing Returns and Artificial Discrimination. Although some of the phenomena mentioned here are not essential to the beginner I thought it would be a good idea, in order that the reader may be aware of the kind of thing that is going on behind his (or her—ED.) back, to include a few details of some of these pretentious devices.

RUMFORD'S PHOTOMETER.

Amongst other things, this instrument is sometimes used to prove that, when light is thrown on a dull subject (as in this book) the angle of ignorance is always equal to the angle of reflection. To use it purely as an anagram machine—as above —is hardly cricket.

Firstly, we have RUMFORD'S PHOTOMETER which is, as you can see, a complicated piece of paraphanalia based on the simple fact that if shadows thrown on the same screen by an opaque body illuminated by two different lights have the same intensity, the illuminating powers of the two lights are equal, if they are at the same distance from the screen, or are in inverse ratio of the squares of these distances, if they are at unequal distances. Next we come to the THEODOLITE, the principle of which is made only too clear in the accompanying illustration.

STARGAZER'S THEODOLITE

This attractive little set up has all the naive charm of the Wimshurst machine, without any of the shocking implications. It is fortunate that the principles it is supposed to demonstrate are so unimportant that the lecturer can soon get down to the more serious business of projecting double-headed rabbits on the screen—using only two fingers and a thumb.

ANAMORPHOSIS is, as one would expect, the opposite to what happens to Butterflies when they emerge from the chrysalis.

THE MEGASCOPE consists of a dark chamber used for the purpose of reproducing an object on a large scale. It would seem that models for this type of work are drawn almost exclusively from the Fakir class, since, in order to reach the posing-platform, a working knowledge of the Indian Rope-Trick is patently required.

THE MEGASCOPE.-*This instrument has the happy knack of turning a subject upside down without disturbing the drapery. For those who hate the usual tomboy tricks of the studio and prefer to work quietly in the dark, what could be nicer ?*

UNUSUAL-VIEWPOINT PHOTOGRAPHY.

It is refreshing, after all the foregoing examples of misapplied ingenuity, to get right back to a few modestly practical ideas of my own ; ideas, I may say, in which the brain, rather than complicated paraphanalia, is the *motor* force. As a demonstration of the way in which quite simple means can be utilized to attain worthy ends I refer the reader to the illustration on page 62, which shows a lively scene at a *The Dansant* taken through the glass bottom of a pewter pot : if I remember rightly the pot contained about half a litre of old-and-bock at the time

Photograph of a Thé Dansant taken through the bottom of a pewter tankard.

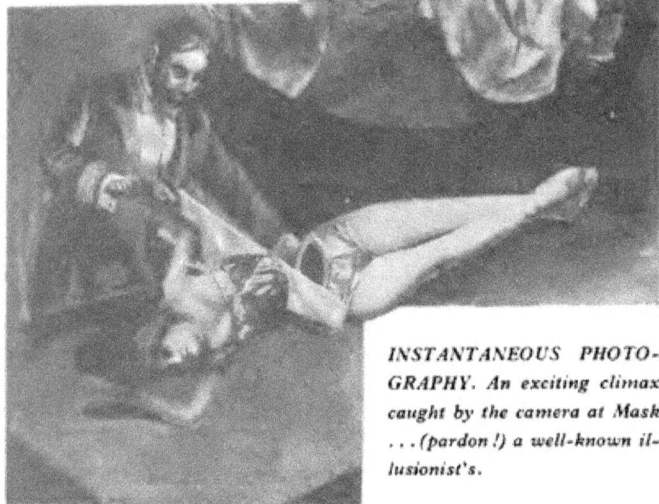

INSTANTANEOUS PHOTO-GRAPHY. An exciting climax caught by the camera at Mask . . . (pardon!) a well-known il-lusionist's.

INSTANTANEOUS PHOTOGRAPHY.

This branch of photographic endeavour calls for nimble fingers and a watchful eye : with these attributes, a hand-camera, and the co-operation of the management almost anyone can take pictures which recall with dramatic intensity those never-to-be-forgotten moments of vaudeville, burlesque, symphony concerts and real life. I can honestly say that my shot of " Sawing the Lady in Half " has done as much as anything else to put this exhilarating pastime on the map. (*See chapter " Moving Objects.*").

TRANSPOSITION.

This aberration has not, as is commonly supposed, anything to do with Buddhism ; but on the now established principle that " *Boy Bites Dog* " is news, whereas " *Dog Bites Boy* " is not, one is surely entitled to take a few pictorial liberties. It is partly because TRANSPOSED-SUBJECT photography is to some extent indicative of the new spirit of healthy scepticism, that is sweeping through the darkrooms of to-day, and partly because it isn't, that I propose to deal with it at some length.

Who has not, in moments of searing vision, ennui, or pique, itched to upset—even if only pictorially—some of the humdrum, established situations of History, Science, Art, Entymology and Domestic Relations ? For instance, one cannot fail to get a little blasé about " Pharaoh's Daughter Finding Moses in the Bulrushes " . . . so why not reverse the situation and introduce new trains of thought by portraying " Moses Finding Pharaoh's Daughter " in similarly wild surroundings, as I have done in my photopicture on page 65.

Naturally, one sometimes makes mistakes ! . . . and although at first I considered " Two Bicycle Maids " (*see photograph on page* 64) to be ethically superior to a " Bicycle Made for Two," I was quick to agree with critics who pointed out that a young woman brazen enough to smoke—even on a bicycle and in the comparative seclusion of a wood—was more likely to be a hussey than a maid. However, the truly enthusiastic photographer soon learns to take bloomers of this sort in his stride.

TRANSPOSITION.—Yet another vivid example of this fascinating art. My "Two Bicycle Maids" has a purer, sweeter significance than "A Bicycle Made for Two" could ever have.

TRANSPOSITION.—*Here we have a clever variation of a hackneyed theme. My photopicture " Moses Finding Pharaoh's Daughter in the Bulrushes " is such an obvious improvement on the original that it would be pointless to dwell on it . . . or would it ?*

BOTTLING.

From *ships*-in-bottles to *people*-in-bottles is but a short step to the enterprising photographer. From the first it is well to realise that some subjects are bottled more easily than others ; careful initial choice can obviate a lot of useless effort. The following illustrations serve better than a spate of words to explain what I mean. In the top picture the subject is easily and comfortably accommodated by quite an ordinary type of bottle and she looks relaxed and pleased with both herself and her surroundings ; whereas, in the bottom right-hand picture, despite a certain attitude of defiance and a rather unusually shaped bottle, the subject is obviously ill at ease and bursting to escape from it all. The other little girl looks happy enough, but the bottle she is in cost more money than, in my opinion, the result was worth. In conclusion I would suggest that subjects for bottling should be acquired nett (top), rather than gross (bottom, right).

Colouring-up and Lantern-Slides

There are some would-be purists who assert that the addition of colour to mono- chrome prints is both unnecessary and inartistic. With this dictum no right- thinking photographer can possibly agree : unnecessary, perhaps ! . . . inartistic, NEVER ! What could be more attractive than a nicely coloured-up print of some loved one . . . or a happy family group . . . or something ? Who, indeed, has not been struck at some time or another with a feeling of acute frustration when, in the course of a pleasant country ramble, armed only with a camera and a stand, one is confronted with some colourful and picturesque scenes such as I have recorded on pages 80 and 81.

No, I regret to have to say it, but the ANTI-COLOURING-UP CAMPAIGN that has swept through the photographic fraternity like a blight is nothing more or less than a vile attempt by vested interests—represented by a handful of unscrupulous R.A.'s—to confine the monopoly of the manufacture of coloured pictures to a small, privileged group. Photographers and the public generally would do well to ignore such obviously biased and defamatory criticism as is so assiduously fostered by this unprincipled and self-seeking minority.

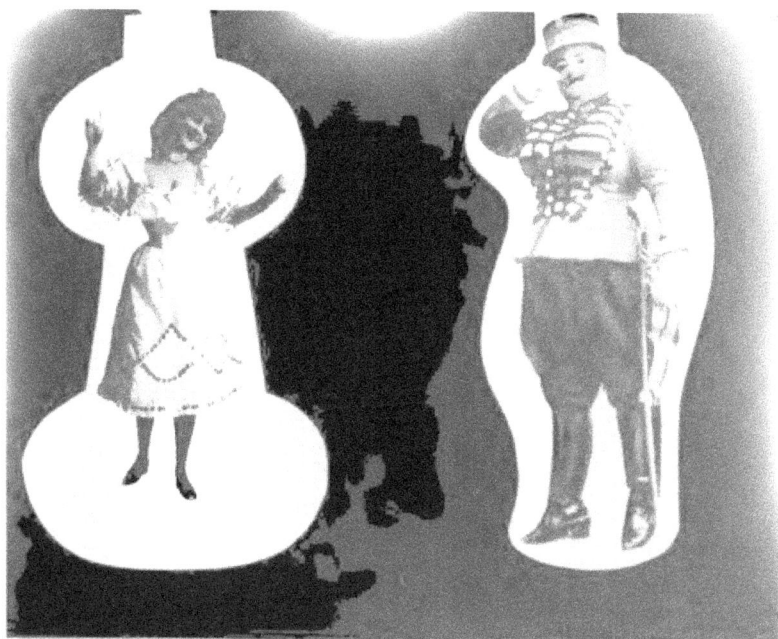

Examples of " Bottling," the fascinating possibilities of which have recently set the photographic world agog. For detailed comment the reader is referred to page 66 of this book and the Ency. Britt.

I will say most definitely that anybody—yes, *anybody*—who is capable of using his (or her—? ED.) eyes and of making a few pencilled notes can decorate an ordinary photograph so effectively that it is quite worthy to rank, in artistic value, with the over-puffed productions of professional contemporary painters. The procedure is roughly as follows : First choose a scene the composition of which is completely in accord with one's finer feelings and the teachings of Mr. Ruskin, and proceed to photograph it in the usual manner. Then remove, develop and fix the plate in a PORTABLE COLLAPSIBLE BAG TENT (U.C.E.) of the type shown in the accompanying illustration.

The fact that the uninitiated never know, and find it difficult to guess, what is going on inside the Bag-Tent, has led—we are sorry to say—to its widespread abuse. Dilletante photographers often impose on the credulous public and make a positive, if somewhat negative, nuisance of themselves by producing nothing but white rabbits and yards of coloured ribbon from its capacious depths.

This ingenuous adaptation of the Carlyle Cloak (*for a description of which see section on Outdoor Photography*), is specifically designed to enable the operator, whilst processing the plate, to peer from time to time at the scenery, etc., and thus facilitate the memorising of the various colours. Incidentally, the hands may be removed at will from the Bag Tent to enable brief notes regarding the colour to be jotted down on the washable celluloid cuffs and dickey without which the well-trained male photographer is seldom to be found. I will not presume to make suggestions to the ladies regarding appropriate places where they could jot down *their* notes.

A word of warning !   Never, when using the cuff-and-dickey method, use indelible pencil or ink : if a permanent record is essential it is far more conducive to domestic harmony if the notes are written lightly in pencil and transcribed into a suitable notebook on arrival at the studio.   To give you the idea here is a facsimile of the original colour notes for my Salon Diploma-Winner, entitled " Les Land Girls " (*see page* 43).

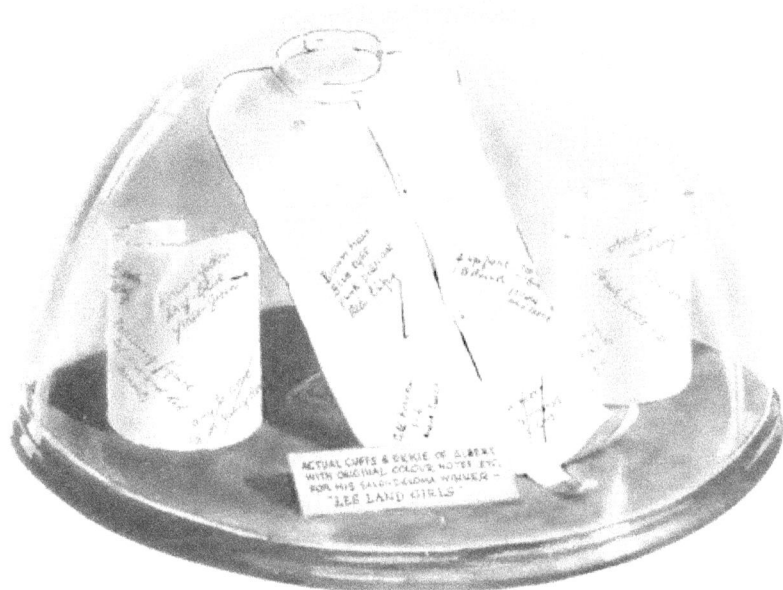

ACTUAL CUFFS & DICKIE OF ALSTAC WITH ORIGINAL COLOUR NOTES ETC FOR HIS SALON-DIPLOMA WINNER — "LES LAND GIRLS"

As you will observe, certain irrelevant scientific data appertaining to other studies happened to be already on the cuffs, but I took the precaution of crossing this out first, having been foxed on previous occasions by cryptic phrases such as . . . " Any to come 5 - each way reclining figure with " red dress and red bonnet " . . . which misled me into losing ten shillings on a horse called Red Riding Hood.

By this time the plate is usually dry enough to be taken away and printed down in the normal way. A matt surfaced paper is the best for colouring as this will take almost any water, oil or spirit bound pigment without cockling, peeling, stretching, or shrinking excessively ; needless to say, a little of all these qualities is a good thing, as they tend to impart that rugged hand-done appearance so beloved of the connoisseur.

THE ODIFEROUS-OIL PROCESS.—" Composition in Smells " is indeed an art in itself. Only after years of almost suffocating experiment was I able to achieve the mastery that was so apparent in my original photopainting entitled " Warm wether is on the whey." Here the characteristic odours of cow, goat, chicken and ducks were offset with just a touch of Jasmine, asafœtida and Icelandic Stoat to produce the haunting blend that was noticed by practically everyone without a cold at the crowded opening of last year's Salon. Unfortunately, the peculiar smell of this monochrome reproduction gives you no idea at all of the original.

LANTERN SLIDES.—Success on this
bye-way of the photographic art comes
only after much experiment and practice. The two little ladies above would
seem to be doing well enough—but
wait until they get on the slope. The
younger lad on the left has come a
purler right at the start.

The type of colour used is a matter of individual preference—there are several brands of ready prepared PHOTO TINTING LIQUIDS on the market, and I expect that most beginners will prefer to use one of these. For those who are a little more ambitious I would suggest the use of oil colours as these have more body and can be made to stand up in ridges just like real oil paintings. Another little discovery of mine (which had not previously occurred, even to Mr. Rimmel) is that if the pigments used are mixed with pleasantly odiferous media—such as lavender oil, frankincense or myrrh—the pictures can be made to smell like herbaceous borders, or an Old English Garden.

Critics have often remarked that my pictures smell, although, in their ignorance, they never seem able to identify any *particular* smell. Simply to say " Mr. So-and-so's pictures smell " is not enough . . . more precise information is required : to be told that the female figures in my photo-painting, entitled " Warm Wether is on the Whey " smell of jasmine behind the ears and seaweed in the corsage *does* help individuals who have not had an opportunity of appreciating the original to get more pleasure out of a mere reproduction, which, even if it does smell, certainly does not smell in the same way. For examples of the purely visual excellence that can be achieved by the oil-paint technique, the reader is referred to the series of pictures starting on page 80, entitled " Hot Feat."

To quote again from Mr. Welford's interesting book—the Hand-Camera Manual —" One of the most attractive uses to which hand-camera shots can be put, is " that of making lantern slides. By this means we can interest our friends and " show them the results of our last holiday trip. (All of them ?—ED.)

" There are two distinct methods of production, one by reduction in the camera " and the other by what is termed contact printing. As the former is used " principally for the larger size negatives, I need only describe the latter. Special " lantern plates are required. The negative is placed in an ordinary printing " frame, and in the dark-room the lantern plate is put film to film with the negative " and the back inserted. Exposure to artificial light is then made and the plate " developed and finished just the same as a negative.

" The result is, of course, a positive print on glass. When dry, a suitable mask " is selected, a covering or protective piece of glass placed over it, and the two " bound together by slips of paper which are sold ready gummed for the purpose."

Never use oil colours for colouring-up lantern slides as the heat of the lamp makes the paint run and the resultant enlarged image bears even less resemblance than usual to the description given by the harassed lecturer. As a specimen of what *can* be done we show the decorative effect achieved by two little ladies setting out on their first unattended lantern slide . . . in this example the features are pleasant enough not to require a mask. *For further examples see Appendix "A."*

Rude Postcards

Whether one uses the word *rude* in the archaic sense of primitive, simple, unsophisticated, in natural state rugged, unimproved, uncivilised, uneducated, roughly made, coarse, artless, or wanting subtlety—or whether one does not—it is still an undisputed fact that as a means of producing RUDE POSTCARDS, the art of photography is on the up. That being so I propose to give a few hints to the beginner to enable him (or her—ED.), to avoid the usual pitfalls.

Firstly, as in painting, it is not a good thing to model one's rudeness too much on the French School . . . there is a delicate *je ne sais quoi* about the Gallic approach that drives the Sturdy British Public right into any odd corner when confronted with a pictorial sample. This sort of solitary ecstasy is against all the principles of ethics and mass production, and is of dubious educational value. The same general criticism applies to the German, Flemish, Italian, Middle-Eastern and Far-Eastern Schools.

No, the only legitimate approach to the problem is to delve right down into the sub-conscious, if necessary until it hurts, until an IDEA is born : for in Rude Postcards the *idea* is the thing, technique is an altogether secondary consideration. Fortunately it is only necessary to do this once. Armed with the right kind of idea the veriest tyro can produce dozens of saleable variations, which, in conjunction with interchangeable captions, can be magnified by permutation and combination into thousands.

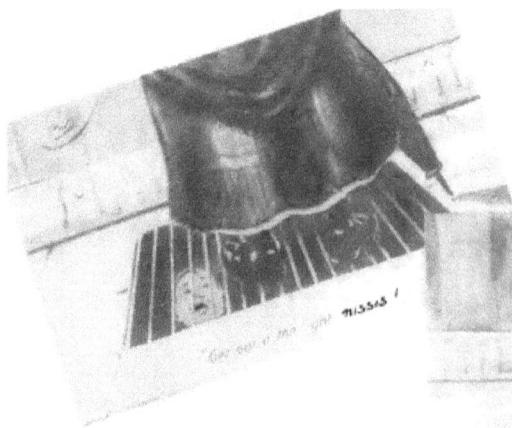

"Get out of my way missis!"

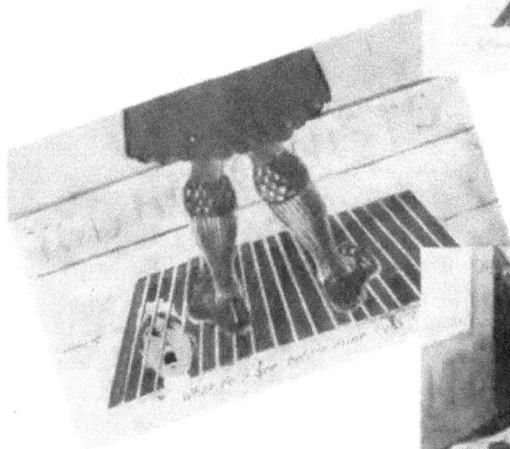

"what goes the postman time"

"what the eye sees
the heart sometimes longs for"

Take the simple idea expressed by the accompanying series : Whilst quietly philosophising in the coal-hole a student has his (or her—Ed.) thoughts and vista of the outside world *rudely* interrupted by a foreign body. Note that expression *rudely* ; it is the operative sentiment and conditions the whole idea, although it is the precise nature of the interrupting body that provides the delicate *nuances* of the variations. Actually the body need not be foreign, although I have usually found that the situation has an added piquancy if it is. Naturally, there are other approaches to the problem, but I can only say that my best Rude-Postcard work has been produced in strict accordance with the above method (*Miss Kelland may know others*—Ed.)

Guide to the Reproductive Processes

Even the few knowledgeable writers on this subject have fallen into the rather tedious habit of handing out a lot of preliminary guff about bees and pollen. One could almost go so far as to say that many a growing lad (or lassie—Ed.) has had his (or her—Ed.) enthusiasm nipped brusquely in the bud by such evasive tactics. I propose to dispense with both bees and pollen—and, indeed, any other red herrings—and get right down to bedrock.

THE PURPOSE OF MECHANICAL REPRODUCTION.

The primary aim is to speed up reproduction so that everybody who wants one can have one. It is erroneous to assume that a reproduction must of necessity be a facsimile of something : modern photo-eugenical reproduction sets out to improve upon the originals.

Print from the YELLOW plate.

Print from the RED plate.

Print from the BLUE plate.

The final print in full colour.

HOW IT IS DONE.

Thoroughly to analyse the various Reproductive Processes is beyond the scope of a purely introductory article of this kind; I will therefore concentrate on broad principles rather than sordid detail.

Briefly, the procedure is as follows: either (1) the original is photographed down on to a suitable printing surface and etched into relief or intaglio, or (2) the original is photographed down on to a suitable printing surface and *not* etched into relief or intaglio.

For example, this book was produced by a special application of the latter principle called Photo-Lithography after a man named Alois Senefelder, who wrote down his greasy washing list on a piece of limestone. Apart from the fact that washing lists are now usually written on the backs of envelopes the process used to-day is very similar.

THE REPRODUCTION OF COLOURED ORIGINALS.

Despite my remarks anent similitude there is one branch of reproduction in which a certain resemblance to the original is almost a social duty. I refer, of course, to *colour* reproduction. If the originals are, for instance, a sort of yellowish-pink, and the reproduction turns out to be a strong chocolate-brown, there is bound to be a lot of local disillusionment and tittle-tattle. For the purpose of reproduction the primary colours are regarded as being Red, Yellow, Blue and Black— not Red, White, and Blue, as is commonly supposed. (N.B.—Printers are the only section of the community to call black a colour.) The colours are usually printed one at a time and compound colours are made by putting the different colours on top of, or very close to, each other—in the same manner as when cheating at Patience.

One way in which this might happen is shown on page 76 (Figs. 1, 2, 3, 4), whilst the series of coloured Continuous Sequence Photographs entitled " Hot Feat " (*see Appendix " A "*) is a good example of the heights to which Photo-Mechanical Colour Reproduction can rise when in the hands of an expert.

Although there is no actual Registrar of Inks and Colours (yet !—ED.), registration is essential. Reproduction without registration has been frowned on for years in the Western World, although it is still encouraged by certain carefree tribes in Bloomsbury and the Upper Congo. The illustration in *Appendix " F "* is a disgusting example of reproduction practised without the slightest regard for registration : one has only to compare it with the legitimate examples on page 80 to see where the difference lies.

In conclusion I would advise beginners to leave reproduction to the experts, for, unless one is constantly aware of a definite urge towards that sort of thing, one soon finds that more time is being spent in worrying about details, errors of omission and commission, and other irritating factors, than can be spared from more exciting photographic pursuits.

Appendix "A"

EXAMPLES OF COLOURED CONTINUOUS SEQUENCE PHOTOGRAPHY.

(1) A SET OF SIX HAND-COLOURED OIL PHOTO-PAINTINGS ENTITLED " HOT FEAT."

I regard this charmingly idyllic sequence as the Photographer's answer to the spasmodic efforts of the new fangled *Bioscopists*. There is no flickering light or jerking movements to worry the eyes and unduly excite the senses ; and each incident can be carefully examined with the leisurely detachment of the student— and in broad daylight, too. Lest it should be thought that the foregoing remarks are prompted by a narrow professional partisanship I hasten to qualify my dislike on ethical grounds :

In my opinion the Bioscope, or Biograph as its sponsors grandiosely call it, clearly represents a vicious attempt to prostitute Art by latently pandering to the low human instinct for gregariousness, even to the extent of downright promiscuity

In " Hot Feat " the plot is simple and direct with no eternal triangles or vicious circles of any kind. The moral is deftly pointed and the colour adds a convincing realism that is vaguely reminiscent of Tintoretto at his best.

The reader will doubtless find it difficult to believe that the basic photographic portions of these superb oil-photo-paintings were taken with an exposure of only one twenty-fourth of a second and an aperture of f8.

(2) A SET OF THIRTY-TWO WATER-COLOURED-UP PHOTOGRAPHS ENTITLED " DON'T BE A FREUD ; HAVE A LOOK ALICE ! "

In this rather ambitious series of pre natal prints it was my intention to present a vivid, visual record of what goes on in the mind of a young student before she does her homework. I would like to take this opportunity of thanking the model— Miss Ophelia Hare-Rhys—for her wholehearted co-operation ; a co-operation without which my intentions could certainly not have been realised.

The outstanding success of this scientific analysis has been generously recognised by most of the leading psychologists of the day, and I have been so inundated with requests for sets of prints, from every conceivable seat of learning, that I am seriously considering taking advertising space in some of the more abstruse scientific journals. A hint that five shillingsworth of stamps should be enclosed with every request, combined with a definite promise that plain envelopes only will be used, will, I think ensure that applications are limited to *bona fide* students.

SPECIAL NOTICE.
For Lecturing and other educational purposes, both " Hot Feat " and " Don't be a Freud ; have a look Alice ! " are obtainable in the form of lantern slides. When ordering please state age next birthday.

(1) " Oh dear, how our poor old feet do burn ! "

(2) " Look Emma, there's a spangle-bottomed beaver hiding in the bushes.

(3) " Ah, that's better, girls—come in before the water boils."

(4) " Such gaiety! The others will never believe us when we tell them."

(5) " Good gracious! And you called him a spangle-bottomed beaver."

(6) " How I wish we'd used Blanks Anti-Foot-burn Ointment instead."

(1) " Oh dear, how I hate Algebra . . .

(2) . . . and Geometry . . .

(3) . . . and History . . .

(4) . . . and Emancipation . . .

(5) ... *and Botany* ...

(6) ... *and Geography* ...

(7) ... *and standing in horse buses* ...

(8) ... *and Economics* ...

84

(9) ...and Literature...

(10) ...and Latin roots...

(11) ...and Gardening...

(12) ...and Cube roots...

(13) ...and Aero-dynamics...

(14) ...and Aspidistras....

(15) ...and Woollen underwear...

(16) ...and Eurythmics...

(17) ... and Woollen underwear ...

(18) ... and Malthus ...

(19) ... and Entymology ...

(20) ... and Spelling ...

(21) ...and Elastic-sided boots...

(22) ...and Woollen underwear...

(23) ...and Cooking...

(24) ...and Chemistry...

(25) ...and Nature Study...

(26) ...and Boyle's Law ...

(27) ...and Geology...

(28) ...and Ballistics ...

(29) ... and the Binomial Theorem ...

(30) ... and Ballet ...

(31) ... and Elastic-sided boots ...

(32) ... and Woollen underwear ...

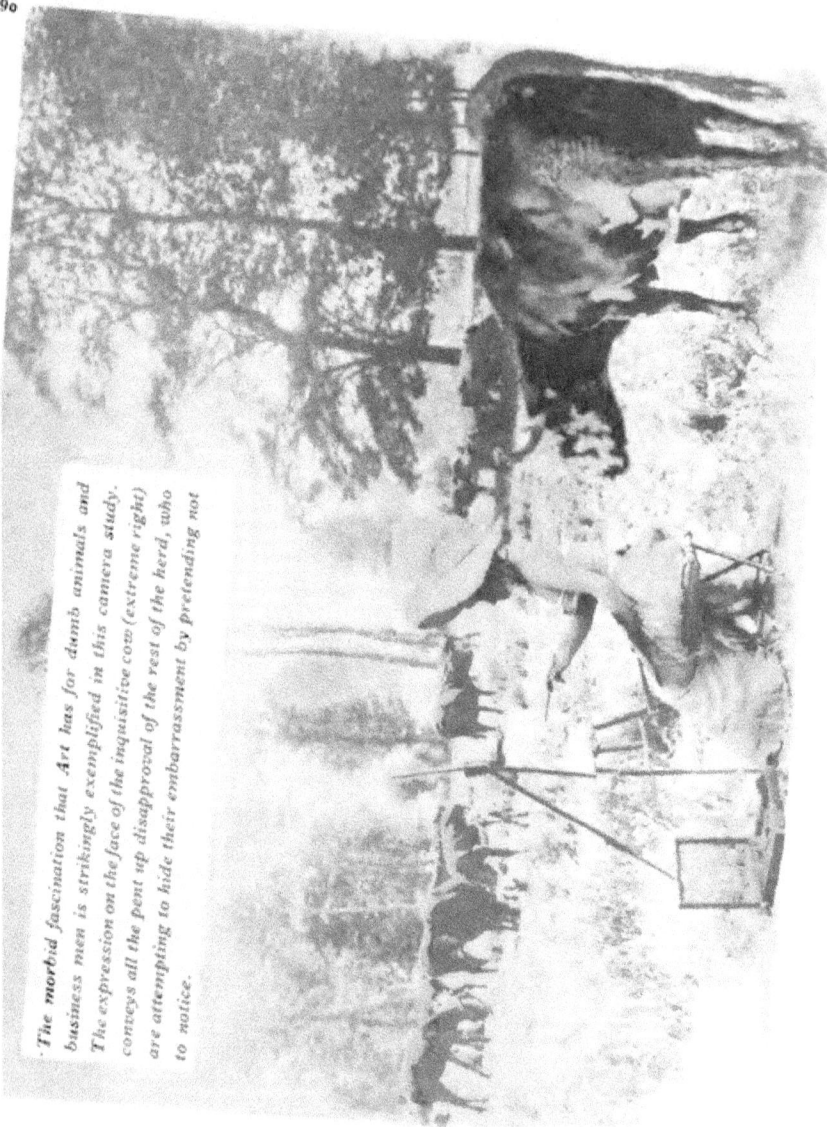

The **morbid** fascination that Art has for dumb animals and business men is strikingly exemplified in this camera study. The expression on the face of the inquisitive cow (extreme right) conveys all the pent up disapproval of the rest of the herd, who are attempting to hide their embarrassment by pretending not to notice.

Appendix "B"

ART IN CAMERA.

All art is relative, but relatives are seldom art. My Aunt Letitia is a case in point : she has a face that terrifies children under twelve, and tradesmen. The latter quality is useful, but you have to live with Aunt Letitia to get the benefit of it : this is unfortunate. Indiarubber is the salvation of Amateur Art, but rubbing-out should not be indulged in except under the direct supervision of an Art master. If you can't get a thing where you want it first time, keep trying on separate pieces of paper.

Great Moderns have been known to draw with burnt match-sticks on the table-cloths in restaurants ; others, under the stress of conflicting emotion, draw on walls in gouache, tempera, charcoal, silver point, pin point and pencil stub. Waiters are unhappy people. To refer to " Hercules Leaning on Club " (see picture on page 42) as " all knobs and bumps " is the height of ignorance—if, after drawing it forty-seven times, you still lack a feeling for form it is because (*a*) you are naturally plump, or (*b*) you had the foresight to bring a cushion. Landseer, G. F. Watts, and a German painter whose name I can't remember, are all great artists—cultured business men bought thousands of their works and gave them to provincial galleries. " The Dobbses are the Medici of the nineteenth century," writes a contemporary critic . . . but I think that Marx is right when he says the Borgias are to blame. Art is a great leveller ; it brings the public down to the level of its most successful exponents.

Drawing as a means of self-expression is better than fretwork, because it is more difficult to put drawings round clocks. To half-close the eyes when looking at a picture is to be a connoisseur—completely to close the eyes when looking at a picture is rude. Any picture is pornographic that has hairs on . . . except Landseer. Blotting paper, rolled up tightly into a pointed cylinder, tones up flabby muscles and reduces observation to a formula.

To copy a photograph at all requires infinite patience ; to copy a photograph so exactly that it is difficult, without a magnifying glass, to tell the copy from the

Following the lead of my brothers-in-art I was persuaded to call this Salon Exhibit
" No. 196." The critics simply loved it, and, exercising ineffable ingenuity, countered
by giving it Balham-wide publicity as "... That brilliant example of modern photo-
graphic art ... dramatising, as it does, a whole vivid chapter of British History ...
will, we are certain, go right down to posterity.... No. 196 or ' Canute had a word
for it, too ' is undoubtedly a masterpiece of the first magnitude."

original is Art. To have studied from the Life is to have lived dangerously ; girls who sit astride are emancipated. Pimples are nature's revenge for being emancipated. Emancipation and emaciation are not necessarily the same thing. Popular Art is very shiny ; this is so that finger marks can be washed off. If Art isn't shiny (a) it isn't art, or (b) it isn't popular. Photographs are not Art unless they are out of focus. There are some ignorant people who can't tell fig-leaves from acanthus leaves. Statues without fig-leaves tend to be porno-graphic . . . except Landseer.

A feeling for drapery is invaluable in a sculptor, painter, or photographer. Etty would have made more money if he'd had more of it. Parts of the human body are beautiful, others are merely functional. Paris capitalizes a low liking for detail. The Greeks had no sense of propriety, but to blame them for it would be priggish because they had no penny post either.

To express an admiration for antiquity is normal; to know anything about it is to be boring. We live in an essentially moral age. Once the Church was Art's greatest patron; now soap is. Cleanliness and Godliness clasp hands across the centuries . . . " Bubbles " symbolizes The Church Triumphant.

All great Art can be useful . . . this is not a wilde statement, young men like Bernard Shaw think so, too. Kipling was right when he said . . . :

> " Creation's cry goes up on high
>
> From age to cheated age :
>
> ' Send us the men who do the work
>
> For which they draw the wage ' . . . "

Kipling is always right. To quote is a sign of erudition; books of quotations are very popular.

To know the name of a picture is more important than knowing what the artist is after; there is a deplorable tendency amongst some modern artists to give their pictures irrelevant names, or, worse still, to give them numbers instead of names. This shows a lack of inventiveness and puts yet another burden on the art critic; naming the picture is a good critic's first job. Many of them are very good at it.

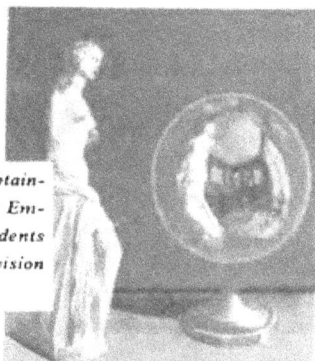

Uncle Albert's Patent Distorting Mirror (obtainable, price 1s. 9d., from all the leading Art Emporiums) is invaluable to those Pure-Art students who do not possess the natural obliquity of vision so necessary in successful modern practice.

Whistler is a great offender; you don't know, without going to the trouble of looking at it, whether " A Study in Grey and Silver " is a review of the Household Troops in Hyde Park, or a portrait of his mother. This sort of thing should be stopped in the interests of popular education. Pictures without names can never be popular. A nice glossy picture with a richly descriptive title makes a good supplement to any periodical. Blatent advertising is when you can read the name on the packet. This is carrying things too far—like putting bladders with words in coming out of angels' mouths.

Patriotic pictures shine up well, but it is better to go back to Agincourt, or further. Otherwise you end up with a canvas full of red coats and brown heathen and the whole thing looks very uneven. Borrowed armour should never be dented, and when painting " King Charles I Saying Farewell to His Children," don't forget the lace collars and the highlights on the curls. Save your studies in case you have the luck to get a commission for a " Blind Boy." Don't try to run before you can walk, but remember, you'll never really get anywhere until you have combined twelve square yards of canvas and a hundredweight of plaster-and-gilt in one picture.

Businessmen like value for money. Love and painting don't mix. No true artist ever loved anybody . . . except Landseer. La Vie de Boheme makes good reading and opera but, unless you have the digestion of a Spanish fly and the scrounging versatility of a Neapolitan urchin, leave it at that. When writing home for money don't mention money or Art . . . talk about " Success being round the corner," " Botticelli's early struggles," and how you miss mother's cake and the old faces. If this doesn't work enclose cuttings from financial papers and underline rising markets in red ink—this shows potential business acumen of a high order.

If you can do a line about a once famous musical comedy star of thirty years back dying in a garret . . . with glimpses of a faded beauty lit with a drink-dazed smile . . . the lights . . . the music . . . the applause . . . and now *this* . . . If you can do a line like this, do it; it's a winner, only make it sordid, or the old man may decide to come and attend to the matter himself; conscience is a fickle jade.

If all else fails buy a plaster skull, place it on a square of black velvet for emphasis,

Students and models relaxing at the Annual Photographic Soiree, recently held at the Albert Hall. Highly technical meetings of this kind provide an invaluable link between the photographer and the model—serving, in a large measure, to break down barriers of reserve built up during working hours.

and concentrate on getting to look like it. Before coma sets in leave a brave note blaming nobody but yourself, turn the gas on in the next room, and drop an aspidestra on the concierge's head. Use French phrases in the note—it shows you've got a Gallic soul. Remember, the average concierge takes fifteen minutes to get up five flights of stairs ; if the aspidestra was a large one it may take him a week. But this should only be used as a last resort and should not be necessary if you move in the Right Circle.

Finding the Right Circle to move in is an artist's first duty. It took Rossetti to turn Giotto's " O " into a circle. Never move in anything but a Circle, it isn't fashionable and, besides, there's an unangular completeness about a Circle that Polygons haven't got. Moving in Circles has all the fluid excitement of Intelligent Discussion ; one knows that, sooner or later, and as inevitably as possible, one will get back to the point from which one started.

The main thing is to have a few people in the Circle from whom one can pick up a little money every time one passes. Opportunism is the soul of *la vie artistique* . . . Montmartre puts a French polish on opportunism.

In Paris a little loose-living is expected, but don't overdo it. Some artists are too loose even for Lautrec. Discretion may be the bitter part of squalor, but without it one's squalor can easily make *assommoir*.

All the best puns are laboured. Work is man's most dignified pursuit next to painting—Ford Madox Brown recognised this. My great aunt Ophelia made seventeen hundred and eighty-three studies from a secondhand plaster cast of Apollo strumming his lyre ; no one knows why she did this. She died a spinster, although she lost caste through studying too long under an energetic, but not very good, painter.

Never suck your brush when doing water-colours ; most water-colours are poisonous. To tell whether a water-colour is poisonous, half close the eyes. Never suck your brush when using oil-colours, it is difficult to get the paint off the teeth. Aunt Ophelia always said that the reason why her teeth dropped out when she was nineteen was because of a misunderstanding—her art master did *not* say " arrange the colours on your *palate* before commencing to paint." Art is a jealous mistress. To tell whether an oil painting is poisonous, half close the eyes.

As an example of loose-fitting rather than close-fitting, I show this candid camera shot of the ... Piecement Artist's and Allied Crafts Biennial ... you cannot avoid comparing the disorderly going-on here with the scholarly calm of the students on page 98.

Appendix "C"

THE POWER OF DOTTED LINES.

Mighty Niagara has already been harnessed to industry . . . the Transatlantic Cable has for years linked continents . . . and of these facts the Great British Public would seem to be both cognisant and appreciative. Is it not then surprising that a scientific discovery of mine—a discovery which will, I humbly prophesy, upset, within a very short space of time, most of our existing notions of Mechanics, Dynamics, Ballistics and Eurythmics—should remain practically unrecognised by contemporary (and if I may say so, *self styled*) leaders of scientific thought.

But perhaps I am not being deliberately cold-shouldered ; perhaps it is the sheer magnitude of my discoveries that makes its realization, even by experts, difficult. Who, indeed, would believe it possible that from the modest beginnings illustrated in Fig. 1, when, grasping a single Dotted-Line in my teeth I bounced an indiarubber ball on one end whilst lifting a sturdy letter A with the other, such amazing feats as those pictured in Figs. 2, 3, 4 and 5 could have developed. I suppose that, under the circumstances, even the Royal Society might be forgiven for being a little incredulous.

As is not infrequently the case the true direction of my endeavours was determined by chance rather than by rigid intention and it was not until I had broken three sets of teeth that I discarded this method. Changing the fulcrum to the top of

Fig. 1.

Fig. 2.

my forehead it was not long before I was able to lift a steel-rule, two M's, two arrows, a P and a Q; using a radial arrangement of Dotted-Lines, as shown in Fig. 2.

I next removed the fulcrum to the left eyeball, and using a heavily tasselled smoking-cap as a counter balance I raised two cigarette cards horizontally in the air with hardly any strain at all on the neck. This is shown in Fig. 3 (the angular piece of tin merely serving to keep the sun out of my eyes).

Fig. 3.

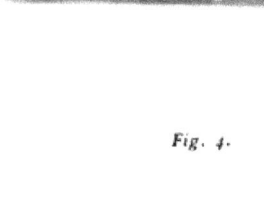

Fig. 4.

Fired by the success of the preceding experiments I next used the right eyeball to project a series of Dotted-Lines through a glass prism and thus cut a lighted candle in half at three feet; the energy released being sufficient to ignite the lower half almost immediately after decapitation. Notice the tensely suggestive attitude of the left hand in Fig. 4.

Even after these impressive examples, Fig. 5 comes as rather a shock—here, with the bridge of the nose as a fulcrum, we have no less than seven snooker balls, two billiard cues and a miscellaneous assortment of letters and figures supported on as complicated a system of Dotted-Lines as one could wish for.

Who can resist a slight feeling of " *Après ca, le déluge* " ?

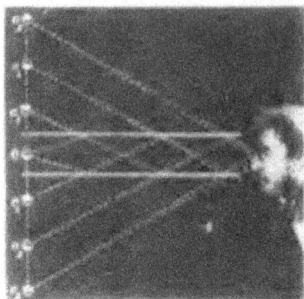

Fig. 5.

Appendix "D"

UNPLEASANT EXAMPLE OF REPRODUCTION WITHOUT REGISTRATION.

www.ingramcontent.com/pod-product-compliance
Lightning Source LLC
Chambersburg PA
CBHW021946190326
41519CB00009B/1151